YOUR KNOWLEDGE HAS VALUE

- We will publish your bachelor's and master's thesis, essays and papers

- Your own eBook and book - sold worldwide in all relevant shops

- Earn money with each sale

Upload your text at www.GRIN.com and publish for free

Johannes Edelhoff

The European Automobile Industry

Technological change, changing of production and changing of organization

GRIN Verlag

Bibliografische Information der Deutschen Nationalbibliothek:

Die Deutsche Bibliothek verzeichnet diese Publikation in der Deutschen National-
bibliografie; detaillierte bibliografische Daten sind im Internet über http://dnb.d-
nb.de/ abrufbar.

Imprint:

Copyright © 2004 GRIN Verlag GmbH
Druck und Bindung: Books on Demand GmbH, Norderstedt Germany
ISBN: 978-3-640-34990-6

This book at GRIN:

http://www.grin.com/en/e-book/128175/the-european-automobile-industry

GRIN - Your knowledge has value

Der GRIN Verlag publiziert seit 1998 wissenschaftliche Arbeiten von Studenten, Hochschullehrern und anderen Akademikern als eBook und gedrucktes Buch. Die Verlagswebsite www.grin.com ist die ideale Plattform zur Veröffentlichung von Hausarbeiten, Abschlussarbeiten, wissenschaftlichen Aufsätzen, Dissertationen und Fachbüchern.

Visit us on the internet:

http://www.grin.com/

http://www.facebook.com/grincom

http://www.twitter.com/grin_com

Universidade de Lisboa

Faculdade de Letras

The European Automobile Industry

Johannes Christian Edelhoff

Geografia
Geografia de Indústria
2003/2004

Contents

1. Introduction

The automobile industry is the key industry in the second part of the twentieth century. Its significance is evident particular in cause of its assemblers. Nowadays 3-4 million people are employed directly in the automobile industry and a further 9-10 million in the manufacture of components and materials. Additional 6 Million people are employed in selling and servicing the vehicles. This leads to number of 20 million employees in this industry (Dicken 1992, p. 268). It is stated as the major industry of the 4th long Kondratief-wave (Blotevogel 2001, p. 203).

Likewise the automobile industry often is mentioned as a prototype of globalized models of production activity and investment. It is one of the first industries to delocalize its activities and become truly "global" (Nunnekamp 2000, p.1).

> "If any sector can be represented the features of globalization it is the automotive industry" (World Trade Agenda 2000, p.1).

As symptoms of this process Vickery (1996) mentions the significance of foreign direct investment (FDI) and production thru subsidiaries companies. He mentions the international network of strategic alliances and collaborations. Production main markets and transnational trade are situated in OECD countries and the 10 largest transnational corporations (TNC) are producing 71% of the world output of vehicles. 14 of the 100 largest companies and 5 of the 10 largest TNC's are automobile producers (Nunnekamp 2000, p.12). In many national economies automobile industry has a major function. Therefore politics of industry and trade play a key role to protect the national automobile industry (Blotevogel 2001, p. 203).

This work focuses on the European automobile industry and attempts to illustrate its advantages and problems. It is divided into 4 sections. First output, trade and market situation of the European automobile industry is outlined. Second changes in production and organisation systems of vehicle producers are discussed. The third part deals with the development of work and jobs in Europe's automobile industry and finally a conclusion and a panorama is tried to be drawn.

2. The European automobile industry in global markets

Roundabout 75% of world automobile production is produced in the Triad. The European market is one of the most important for car sales and car production. Since the 1970s big changes in European automobile production have been happened.

2.1 Output

During the last 50 years the automobile industry has increased in its world production immensely. World production of cars rose from 13.3 million in 1960 up to 36.1 million in 1995 (Vale 1999, p. 115). Automobile production in Europe increased from 5,34 million in 1960 up 14,19 million in 2003[1].

The evolution of automobile industry can be divided into three phases. In the first phase American producers dominated the world production until the 70s. From the 70s up to the beginning of the 80s a bipolarization was developed between American and European car producers. During the 80s Japanese automobile industry gained power and forced a tripolarization of the world automobile market (Vale 1999, p.114) and until now a domination of the Triad in automobile industry can be recognized.

In Europe France and Germany are the dominating producers since 1960. In 1995 Germany produced 13% of worlds automobile production, which is almost equal to its share in 1960 (14%). Between 1960 and 1989 German automobile industry grew with average annual change of 5,2%. France remained its output with a share around 9% from 1960 until 1998. Specific is the decline of United Kingdom's car producers who had a share of 10.4% in 1960 and a share 3.7% in 1989. Even the absolute production did not raise in that time the United Kingdom. A remarkable growth took place in Spain. In 1960 Spain produced circa 43.000 cars and in 1989 1.6 million cars were produced there (Dickens, p. 272). In 1995 Spain's output was 1.9 million (Vale 1999, p. 115) and in Table 1b it can be seen that its output is even larger in 1998. United Kingdoms automobile producers had a total breakdown in the 1970s and a rebuilding of its automobile production system during the 1980s by Japanese transplants. Spain gained share of world automobile production when European companies expanded production into it. In the United Kingdom and in Spain nowadays the production is strongly export orientated towards European market (Hudson & Schamp 1995, p. 221). In 2002 European Unions automobile production fell by 1% in comparison to 2001. Only new construction places Portugal, Spain and the United Kingdom raised their production in that period. As can been seen in table 2 production of all other EU-15 countries dropped in during this time.

Table 1a Share of world automobile production by major producing countries, 1980-1989

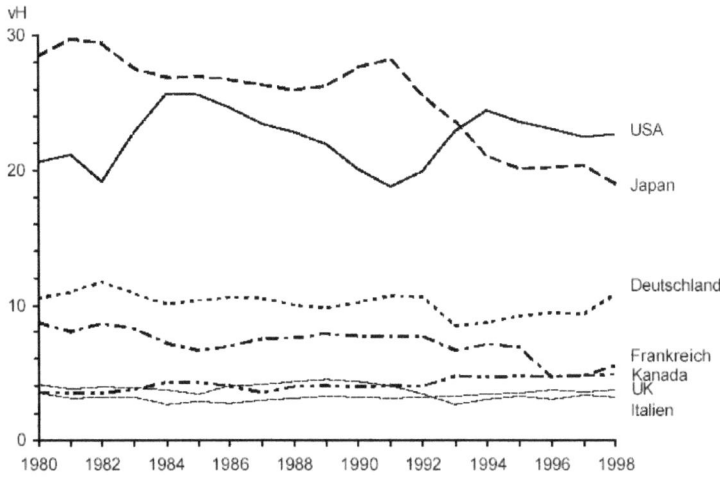

Source: Nunnekamp 2000, Globalisierung der Automobilindustrie: Neue Standorte auf dem Vormarsch, traditionelle Anbieter unter Druck?. Kieler Arbeitspapier No.1002. Kiel. Table 7a

Table 1b Share of world automobile production by new producing countries 1980-1998

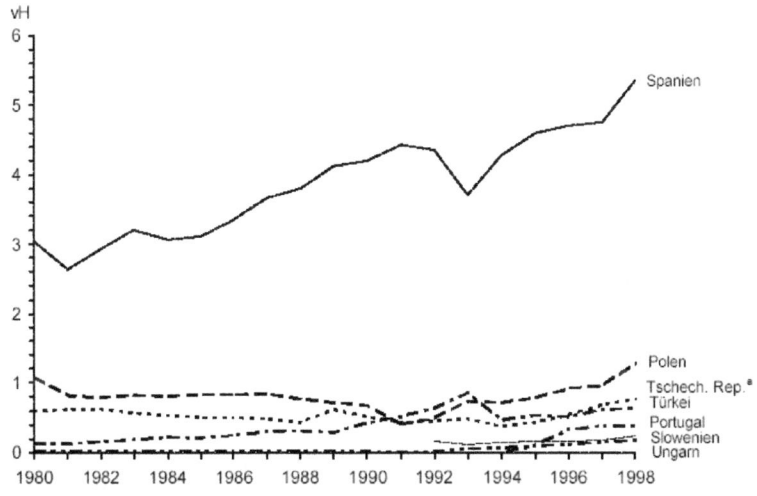

Source: Nunnekamp 2000, Globalisierung der Automobilindustrie: Neue Standorte auf dem Vormarsch, traditionelle Anbieter unter Druck?. Kieler Arbeitspapier No.1002. Kiel. Table 7b

---

[1] Frankfurter Allgemeine Sonntagszeitung, 04.01.2004

Table 2: Motor vehicle production by country 2001 –2002

IN UNITS

CARS	2001	2002	% CHANGE
EUROPE	17 373 368	17 312 114	0%
- EUROPEAN UNION	14 938 604	14 815 406	-1%
Double Countings Germany / Belgium	328 936	297 595	-10%
Double Countings Germany / Austria	39 838	27 000	-32%
AUSTRIA	131 098	131 411	0%
BELGIUM	1 058 656	936 903	-12%
DANEMARK			
FINLAND	41 916	41 068	-2%
FRANCE (1)	3 181 549	3 283 775	3%
GERMANY(2)	5 301 189	5 123 238	-3%
ITALY	1 271 780	1 125 768	-11%
NETHERLANDS(1)	189 261	182 368	-4%
PORTUGAL	177 357	182 573	3%
SPAIN	2 211 172	2 266 902	3%
SWEDEN (3)	251 035	237 975	-5%
UNITED KINGDOM (1)	1 492 365	1 628 020	9%
SWITZERLAND			

Source: OICA

2.2 Global trade and its affect on Europe

A considerable amount of European vehicle production is exported, even as a huge part of
Europeans single markets demand is covered by imports. Trade between European especially
between European Unions countries usually is larger than trade between European countries
and the rest of the world. Almost three-quarters of Western European automobile trade are
intra-regional. Altogether European Unions automobile industry made a surplus of 9,1% in
2002 exports-imports (Sura 2003, p. 6). As Table 3 shows there are enormous differences in
the ratio of imports and exports between individual European countries. Sweden and
Germany have surplus of 33,4% and 32,5%. Also Spain and France have a surplus each
around 13% whereas Italy and the United Kingdom have a shortfall of 40,8% and 37,8%.
These shortfalls and surpluses are the result of complex strategies of TNC's with European
production base. Nearly all inflow of automobiles to Europe is from Asia but the surplus of
9,1% is equalized by an outflow of luxury cars to North America. Japan has not only a trade
surplus towards Europe but also towards North America. Today Europe is an important
market for Japanese exports. The export of Japanese cars to Europe and North America has
been heavily restricted by so-called voluntary export restraint agreements (Dicken et al. 1995,
p.4s). This leads as well to the point that the current tendency toward internationalisation
differs from previous phases. Additional to exports and imports of automobiles FTI has
become more and more important when production is shifted into other countries
(Nunnekamp 2000, p.7). Not only voluntary export restraint agreements led to the raising of
FTI. Furthermore changing in pattern of demand for automobiles have a heavy influence on
the strategies of automobile producers. In 2002 Japanese and Asian cars have been very

successful on European market and have won new shares. For example Toyota, Suzuki, Hyundai and Kia had each a surplus of over 8% on the German market[2]. Today 64% of.

Table 2: Import and Export of automobiles in % of Export- Import in the EU in2002	
Belgium	3,9
Denmark	-54,2
France	13,2
Germany	32,5
Spain	13,7
Netherlands	-12,6
Ireland	-72,8
Italy	-40,8
Luxemburg	-51,1
Austria	1,7
Portugal	-6,4
Finland	-32,8
UK	-37,8
Sweden	33,4
EU 15	9,1

Source: Eurostat

European automobile managers think that today's market share will remain for the coming years[3].

2.3 The influence of demand on automobile industry

Automobile industry is traditionally strongly market oriented. Therefore the development of automobile industry took place in early well-developed costumer markets, which have been North America, Western Europe and Japan (Dicken 1992, p. 278)

Demand for automobiles is sensitive to business cycles. Periods of raising and high demand are separated by periods of dropping and stagnant demand. Strong declines happened in 1982-84 and 1990-1993. Phases of fast growth occurred in 1985-1988 and in 1995-1997 (Blotevogel 2001, p.204). Effect of business cycles on car sales is extremely large. In 1999 40% more vehicles have been sold than in crises year 1993 in European Union (without Germany) (Weiss 2000). Another decline of Europeans car production arises until 2003 when car sales reached a level of 14,19 million. In 2001 14,85 million cars and in 2002 14,42 million cars were sold. Actual prognoses say that car sales will increase in 2004 and 2005 to 15,00 million[4].

---

[2] Süddeutsche Zeitung, 12.5.2003
[3] PS Automobilreport, 11.2003
[4] Frankfurter Allgemeine Sonntagszeitung, 04.01.2004

Rise of motorization in OECD countries led to matured markets, which is a so-called saturation of the market. 85% of demand in these countries is a replacement demand. Markets with a replacement demand are a much slower growing then markets with a new demand.

A heavy problem in world automobile industry and especially in Europe is overproduction. Table 4 shows that Europe had in 2000 an overcapacity of 65000000 units. The utilization rate of manufacturing facilities has decreased from 80 to 69% in recent 10 years of growing overproduction. Firms have to consolidate to solve this problem. The overcapacity of world car manufacturing industry is estimated at 26,000,000 cars per annum, which is equivalent to

Table 4 Production overcapacity (units 000)

Regions	Production Capacity 2000	Demand 2000	Overcapacity 000'	% of demand
North America	22 000	16 000	6 000	38
West Europe	20 500	14 000	6 500	46
Japan	12 500	7 000	5 500	79
Korea	4 000	1 500	2 500	167
Asia	8 000	5 000	3 000	60
Latin America	4 500	3 500	1 000	29
CEE/CIS	4 500	3 500	1 000	29
Others	4 000	3 500	500	14
Total	80 000	54 000	26 000	48

Sources: J.D. Power-LMC Forecasting Service, Salomon Brothers Inc./Smith Barney Inc.

a capacity of about 100 assembling plants. Only the emerging markets with growing consumers' demand are able to recuperate the utilization rate of the world automotive industry in the long run.

Furthermore replacement demand depends stronger on business cycles. In times of recession consumers postpone their demand to a later date. Automobile firms try to solve this problem by developing more different models and by stylising demand. This often produces only marginal changes in automobile models like in the fashion industry.

On the one hand domestic markets are very homogeny. In North America is a demand in large, comfortable cars in Europe demand is in highly affluent, highly mobile, cheap- energy based cars. On the other hand exists a current tendency towards segmentation of the market into different types like Vans, City-Cars, 4wd Jeep et cetera that means a greater variety of cars (Dicken 1992, p.278s). As a result of those marketing strategies consumers in OECD countries lower their using permanence of cars and change it more often (Blotevogel 2001, p.205). The average life cycle of a European car decreased from nine years in 1990 to seven years in 1999[5]. Although automobile industry seems to be very global, regional markets are

---

[5] Magna Steyr, 2002

diversified on the one hand by variations of consumer preferences and household income levels and on the other hand by state regulatory networks which affect car consumption. In European country local brands and producers remain market leaders in their national markets (Hudson & Schamp 1995, p220). In 2003 Volkswagen, Mercedes and Opel are German markets leaders.

Change of demand, overcapacity, slower growth ratings and a more diversified demand, requires new strategies of production and structure of automobile industry and producers have to proof their adaptability.

3. Technological change, changing of production and changing of organization of production

In organization of production exist three big different production systems. Craft production, which was the first way of producing cars mass production, which was invented by Henry Ford, and lean production which origins from Japan.

3.1 From craft Production to mass production

As mentioned car production took place in consumer markets that high levels of demand have permitted the achievement of economies of scale. Therefore mass production was head and shoulders above craft production in offering affordable cars. The most important difference between craft production and mass production is production volume and costs of production.

In craft production unstandardized components are installed in succession and highly skilled workers are needed. Quality of cars are directly linked to workers skills and increasing more cars does not drop production costs. Even today some European producers, that sell their products in market niches, are using craft production like Aston Martin and Rolls Royce (Vale 1995, p.103s) but even Rolls Royce does not use own engines but BMW engines (Martin & Schumann 1996, p.176).

In 1913 Henry Ford introduced the moving assembly line into automobile production, which guaranteed extremely high output within low cost. In following years until the 1970s automobile production was the mass production industry. In mass production assemblers geographically can be narrowed far away because of a just-in-case system of producers. This assures low prices for components. Workers are no more than low skilled and high specialized. Production of an automobile from design to marketing is a very cost intensive process. It cost's up to 3 billion dollars. For that reason only very large scale models produced in mass production achieved profit. Opinion was shared that 2 million vehicles per year had to be produced to reach the maximum benefits from scale economies (Dicken 1992, p. 279s).

3.2. Lean Production in Europe

At the beginning of the 1980s European car producers had to face new major challenges. Not only the matured markets, which were characterized by slow growing demand and overcapacity, caused new challenges. Also the Japanese challenge appeared. Japanese automobile producers were able to produce new sectoral efficiency norms with higher quality and lower cost and prices (Savary 1995, p. 148). The new forms of organization structure initiated by Toyota are characterized as lean production. Its characteristics can be seen in Table 5.

Lean production is strongly linked to new market demands and new market strategies because it makes shorter design periods per model possible. Although new cars are always more complex than older ones automobile producers try to simplify their production and concentrate on assembling the entire car as a series of sub-assembling. This means that suppliers develop parts of the car (Dicken 1992, p. 281s). Trough lean production automobile producers try to connect advantages of mass production with consumer oriented and quality oriented production of small scales. Therefore it is not the end but a changing of mass production.

Confronted with the Japanese challenge and a declining market major European automobile groups [Peugeot, Fiat, Volkswagen, Volvo, Rover, Mercedes-.Benz (today Daimler Chrysler)] adapted their production and organisation to lean production. This includes a decrease of time taking in designing a new model, a faster and better internal communication, a quick adaptation to consumer demand, a improved product quality at every stage of the value chain, introducing just-in-time production, a reduce of costs through by automating production and a new supplier structure with fewer and more competitive suppliers (Savary 1995, p.149).

Although lean production seems to be very successful tendencies can be obeyed that European car producers disassociate themselves from lean production. In opinion of Springer (2000) companies doubt the success of lean production and try to find new solutions. In his eyes a re- standardization happens in European automobile branch.

3.3 Strategic Alliances and collaborations

Research and developing (R&D) have become one of the main costs of automobile production. Between 1993 and 2003 R&D investments in European's automobile industry increased from 10.4 billion Euros to 18.2 billion Euros[6]. Saving money in this sector is important for every single producer. One possibility to reduce R&D costs is platform consolidation like Volkswagen group does. Volkswagen uses identical platforms for its Seat, Skoda, Audi and Volkswagen models. The reason for using the same platform is that companies strive to reduce the time and effort they spend in developing a new model (Vickery 1996, p. 191).

For the reason of R&D costs producers are creating strategic alliance and collaborations to develop new techniques or new components. Collaborations, acquisitions and alliances have four main objectives. Companies want to penetrate new markets, to improve efficiency of production, to increase the variety of models and to gain innovations (Vale 1999, p. 179s).

A new dimension of networks and simplification frees labour and production across boarders of countries and also across boarders of companies. German cars only exist in luxury class. The 1995 Polo of Volkswagen produced in Wolfsburg is manufactured by a rate of almost 50% in foreign countries. The list of supplier countries ranges from the Czech Republic over Italy, Spain and France to Mexico and the United States. Toyota produces more in overseas countries than in Japan and American automobile producers would break down without Japanese suppliers. Not only the marking 'Made in Germany' also 'Made by Mercedes' or 'Made by Ford' misleads. Automobile producers discovered alliances to share components production. Volvo uses Diesel engines of Audi produced in Hungary and Mercedes buys its engine for Mercedes Viano at Volkswagen. At the same time producers are constructing alliances joint ventures and acquisitions (Martin & Schumann 1995, p. 175).

Until 1980 only 27 co-operation contracts were signed. At the end of the 1980s the number raised up to 70 new contracts and in the time form 1990-92 29 new contracts were signed (Vale 1999, p. 179s). In the early 1990s an uncountable network of partnerships alliances and joint ventures arose and in the late 1990s the number of transactions increased. The monumentum of record deal activity in the automotive sector in1998 when Daimler Chrysler merged continued in the following years. Without the record deal of Daimler Chrysler in 1998 the average volume of transaction grew by 80% up to 275 million Dollars in 1999.

---

[6] ZEW, Mannheimer Innovationspanel, 2002

Since 1998 General Motors and Ford purchased parts of Saab and Volvo. German companies tried to acquire British firms (BMW/Rover) and there have been European joint ventures between Ford and Peugeot Citroen (Nunnekamp 2000, p.9).

Volkswagen and Ford have founded the AutoEuropa Project in Setubal, Portugal. Three major tendencies in automobile production can be noticed in this project. First tendency is to reduce conventional production costs mainly through low wages and states' subsidies in a politically stable environment with an access to a huge consumer market, which is in the case the more European Unions market than the Portuguese (Ferrão & Vale 1995, p.205s). In 2001 98% of AutoEuropas production is exported which 11% of total Portuguese exports.[7] Second tendency is the production of more or less equal cars (Ford Galaxy/ VW Sharan) by two different car producers. R&D costs are reduced immensely through this behaviour and producers can offer a large variety of different car types (Martin & Schumann 1995, p.176). This leads to the third tendency of today automobile market especially in Europe. Special car types, like here multi-purpose vehicles have to been offered to stand competition. Only through reducing of costs a large variety can be guaranteed.

3.4 Greenfield Investment

The case of AutoEuropa is like Nissan in the Northeast region in the UK an example for a Greenfield investment strategy in European periphery. Government are often willing to give high promotion to industrial policy to seek investment, which offers jobs and knowledge. Greenfield investments seem to bring a demonstration effect for surrounding areas. Greenfield investments seem to have several advantages like low corporate and personal taxation rates, low labour cost and social cost a deregulated labour utilisation, a entrance to European Unions' market, high labour quality and flexibility and mostly high government promotions. Therefore huge parts of Greenfield projects are often paid by the government and so by tax payers (Pike & Vale 1996, p.98s).

A problem about Greenfield investment is its high investments into new production facilities and infrastructure increase expenditure in proportion of output growth. Automobile industry has an enormous overcapacity and its sales depend strongly on economic cycles. In Greenfield plants job growth is often smaller than announced before because production never will use its full capacity:

> "The initially stated objects regarding employment and production were not achieved.
> In fact, the forecast creation of 5,000 jobs generated by the company and a production

---

[7] Frankfurt Allgemeine Zeitung, 28.01.2001

output of 180,000 vehicles a year were not achieved (the volume was only around 50,000 vehicles in 1995, although only by the end of the decade the production should be reached" (Pike & Vale 1995, p.112).

Even in 2001 the full capacity was not achieved. The output was 580 vehicles a day when capacity was 800 vehicles a day. The future of AutoEuropa is not secure because at the end of 2004 production of Galaxy and Sharan will probably stopped[8].

An important limitation for Greenfield strategies is time. Greenfield plants age and mature and loose their former advantages. AutoEuropa will have problems in gaining new orders on account of the enlargement of the European Union toward East (Pike & Vale, p.112s.)

4. Jobs in automobile Industry

Changes in demand for automobiles, changes in structure and organization of process of production, new networks of suppliers, consolidation and overcapacity forced European automobile producers to do changes in employment structure. Labour forces have been reduced immensely during the 1980s until today. During the change from Fordistic mass production to lean production at first low skilled workers and mass production workers became unemployed:

> "the current Western automobile workforce is in precisely the opposite position of craft workers in1913. The introduction of mass production created new jobs for craft workers – these workers made the production tools needed by the new system. By contrast, lean production displaces armies of mass-production workers who by the nature of this system have no skills and no place to go" (Womack, Jones and Roos, 1990,p. 235-6, quoted from Dicken 1992, p.308).

4.1. Labour force

In the 1980s great rationalism took place in Europe's automobile plants. The United Kingdom lost 34,92% of its jobs in automobile production, which is a reduce of 120.500 jobs. The increase of employment created by Greenfield investments in the United Kingdom did not succeed to compensate the degeneration elsewhere. Even more the new unemployed old car workers will hardly find a new job (Pike & Vale 1996, p. 108 s.).

Ford reduced its European labour force as well as Renault in the 1980s whose work force declined from 89.000 to 70.000. PSA let 22% of their working staff into unemployment and Saab cut of its personnel from 17.000 to 11.000 during the same period (Dicken 1992, p. 306).

---

[8] Frankfurt Allgemeine Zeitung, 28.02.2001

In reaction of heavy difficulties Fiat started 1980 a cost-cutting policy. Over 40% of its work force was sent into unemployment. The number of workers was reduced from 134.621 to 77.910 in 1986. The productivity levels of workers increased extremely from 14,8 cars/worker in 1979 to 27,9 cars/worker in 1985(Conti & Enrietti 1995, p.132).

In opposite to the decline of employment in automobile industry in other European countries the work force in car manufacturing increased in Germany during the 1980s. Growth succeeded through enlarging variety of components and parts and through successful policies with labour unions. But also German automobile industry had to force its reconstruction during the crises in the early 1990s. Mercedes Benz reduced 25.000 jobs, VW 13.000 and Opel 6.000. Things have changed during that time in public debate. For trade unions it was more evident to rescue a whole plant and not only jobs (Schamp 1995, S.112s.). Martin and Schumann worked out a cut off of 300.000 jobs between 1991 and 1995 in German's automobile industry. Europeans Ford chef Albert Caspers dropped the productivity levels from 1995 to 2000 from 30 hours per car to 17,5 hours per car. Volkswagen let 7000- 8000 workers each year to unemployment during the 1990s. Output of the German industry did not fall at the same time.

Between 1995 and 2000 the labour force in the European Union nearly stayed equal (-0,1%) In France (-3,9%), Italy (-5,3%) and the United Kingdom (-2,3%) labour force was reduced whereas Spain (2,0%), Germany (2,2%) and Portugal (5,7%) had a surplus in labour force.

4.2. Structure of labour

According of the decline of labour force today the European automobile industry has reached a high productivity level. Its productivity level is 20% more efficient than the rest of the processing industries. Also the education of workers is superior to other processing industries. The lowest education level of workers has Italy. High productivity level and high education lead to high wages in automobile industry. In European Union wages are 25% higher than in other processing industries (Sura 2000, p.3-6).

High productivity level in automobile production is reached through new organisation of work. Working time is more and more disconnected from productivity time. Working shifts have been made for flexible. Working system changed from two-shift system to three-shift system with night shift. Producing time with a higher capital intensity has been reached without enlarging work time. Producing in a two-shift system is disadvantage in today's automobile industry and seems to be more important than total working hours. Volkswagen Wolfsburg reduced working time from 1615 hours per year (1990) to 1560 hours per year

(1998) and increased production time at the time from 3380 hours per year to 4975 hours per year. In times of weak demand factories can go back to two-shift system. This results a more flexible production system while less factories are needed for production. For that reason automobile companies with a huge overcapacity are able to close down factories like Renault did in Belgium and Ford in the United Kingdom (Lehndorff 2000, p.44s).

Conclusion

This work has drawn a picture of Europe's automobile structure and its problems. The Output of Europeans automobile factories reached after a long period of growth a steady level of production. A immense growth for new capacities is not to expect because today there is a immense overcapacity in Europe. During the 1970s and 1980s when Japanese car producers penetrated the European market automobile producers had to start a developing process into their structure, organisation and production system, which has not been finished yet. Even today Asian car are gaining new shares of Europeans vehicle market which is challenged to replacement demand. A bigger variety of vehicles and new flexible production systems which can react on business cycles are already installed and have to be improved if European companies want to remain in global challenge. During the last decades lean production, consolidation and strategic alliances have created a new situation on the European and on the World car market. Only car producers in niches (Alfa Romeo, Porsche) and global giant companies with global strategies are going to survive this competition like today GM-Opel-Fiat-Subaru, Ford-Mazda-Daewoo, Daimler-Chrysler-Mitsubishi-Hyundai, Volkswagen-Seat-Skoda and Toyota.

In the restructuring phase of automobile industry a immense number of employees became unemployed. Today's market seems to remain at a steady number of employees but high overcapacity and neoclassical rationalism are great dangers for European employment markets. 60 % of Europe's automobile managers think that more consolidations are taking place in the following years[9], which is another danger of new unemployment. In today's factories workers have to better educated and more flexible than in earlier times. Labour Unions seem to loose influence during this process, which is a tendency not only in automobile industry.

With the enlargement of the European Union towards East- Europe a new competition in Europe can be started. Wages in East Europe are low and the quality of work is quite good. In

---

[9] PS Automobilreport, 11.2003

neoclassical logic a bigger market of educated employees could lower wages even in Western European factories.

Bibliography

Blotevogel, H. H. (2001): Sektorale Betrachtung III: Automobilindustrie. in: Blotevogel. Weltwirtschaftsgeographie. Berlin: 203- 215

Conti, S., Enrietti, A. (1995): The Italian automobile industry and the case of Fiat: one country, one company, one market? in: Hudson, R., Schamp, E.W. (eds.): towards a new map of automobile manufacturing in Europe. new production concepts and spatial restructuring. Berlin:117- 146

Dicken, P. (1992): Global Shift: the internationalization of economic activity. 2nd. Ed.. London

Dicken, P., Hudson, R., Schamp, E.W. (1995): New challenges to the automobile production system in Europe. in: Hudson, R., Schamp, E.W. (eds.): towards a new map of automobile manufacturing in Europe. new production concepts and spatial restructuring. Berlin: 1-20

Ferrão, J., Vale, M. (1995): Multi-purpose vehicles, a new opportunity for the periphery? Lessons from the Ford\VW Project (Portugal). in: Hudson, R., Schamp, E.W. (eds.): towards a new map of automobile manufacturing in Europe. new production concepts and spatial restructuring. Berlin: 195- 218

Hudson, R., Schamp, E.W. (1995): Interdependent and uneven development in spatial reorganisation of the automobile production systems in Europe. in: Hudson, R., Schamp, E.W. (eds.): towards a new map of automobile manufacturing in Europe. new production concepts and spatial restructuring. Berlin: 219- 243

Lehndorff, S. (2000): Die Arbeits- und Betriebszeiten in der europäischen Automobilindustrie.Gelsenkirchen

Martin, H., Schumann H. (1995): Die Globalisierungsfalle, der Angriff auf Demokratie und Wohlstand. Hamburg

Nunnekamp, P. (2000): Globalisierung der Automobilindustrie: Neue Standorte auf dem Vormarsch, traditionelle Anbieter unter Druck?. Kieler Arbeitspapier No.1002. Kiel

Pike, A., Vale, M. (1996): 'Greenfields' and 'Brownfields': automotive industrial developement in the UK and in Portugal. in: Finisterra, XXXI, 62, 1996: 97- 119

Savary, J. (1995): Competitive strategies in the world market: the case of Renault and the emergence of a European group? in: Hudson, R., Schamp, E.W. (eds.): towards a new map of automobile manufacturing in Europe. new production concepts and spatial restructuring. Berlin: 147- 172

Schamp, E. (1995): The German automobile production system going Europe. in: Hudson, R., Schamp, E.W. (eds.): towards a new map of automobile manufacturing in Europe. new production concepts and spatial restructuring. Berlin: 93- 116

Springer, R. (2000): Rückkehr zum Taylorismus? Arbeitspolitik in der Automobil- Industrie am Scheideweg. Berlin

Sura, W. (2003): Fahrzeugbau in der EU. in: Eurostat: Statistik kurz gefasst, Industrie, Handel und Dienstleistungen, Thema 4 – 28/2003.

Vale, M. (1999): Geografia da indústria automóvel num contexto de globalização, imbricação espacial do sistema Autoeuropa. Disertação de Doutoramento em Geografia Humana, Universidade de Lisboa. Lisbon

Vickery, G. (1996): Globalisation in the automobile industry. in: OECD, globalisation of industry. Overview and sector reports. Paris: 153- 205

Weiß, J. (2000). Die deutsche Automobilindustrie im internationalen Wettbewerb. DIW-Wochenbericht 12/00; http://www.diw-berlin.de/diwwbd/00-12-2.html

World Trade Agenda (2000): Global or Not: the auto sector looks open to new trade disputes and heavy pressure on market opening and investment terms. No.00/01. Geneve